Reading Essentials
in Science

LIVING WONDERS

Cells

SUSAN GLASS

PERFECTION LEARNING

Editorial Director: Susan C. Thies
Editor: Mary L. Bush
Design Director: Randy Messer
Book Design: Michelle Glass
Cover Design: Michael A. Aspengren

A special thanks to the following for his scientific review of the book:

Paul Pistek, Instructor of Biological Sciences, North Iowa Area Community College

Image Credits:

Gamma Ray Studio/Getty Images: p. 34; ©Reuters NewMedia Inc./CORBIS: p. 6 (far right);
©Clouds Hill Imaging Ltd./CORBIS: p. 22 (top); ©Jim Zuckerman/CORBIS: pp. 25, 26 (top);
©Ron Boardman/Rank Lane Picture Agency/CORBIS: p. 29; ©Gabe Palmer/CORBIS:
p. 33; ©Charles O'Rear/CORBIS: p. 35

Photos.com: front cover, back cover, pp. 1, 2–3, 4, 5, 8, 15, 16, 18, 19, 20 (bottom),
22 (bottom), 24, 26 (bottom), 30, 31, 36, 37, 38–39, 40, 41, 42, 43, 46, 47,
background on all spreads, background behind all sidebars; Ingram Publishing:
pp. 17, 20 (top); LifeART®/Lippincott Williams & Wilkins: pp. 26 (middle diagram),
28, 32; ©Royalty-Free/CORBIS: p. 28; ArtToday (arttoday.com):
front cover (right-front), pp. 6 (line art), 7, 9, 14, 27;
Michael A. Aspengren: pp. 10, 12

For information, contact
Perfection Learning® Corporation
1000 North Second Avenue, P.O. Box 500
Logan, Iowa 51546-0500.
Phone: 1-800-831-4190
Fax: 1-800-543-2745
perfectionlearning.com

1 2 3 4 5 6 PP 09 08 07 06 05 04
ISBN 0-7891-6227-x

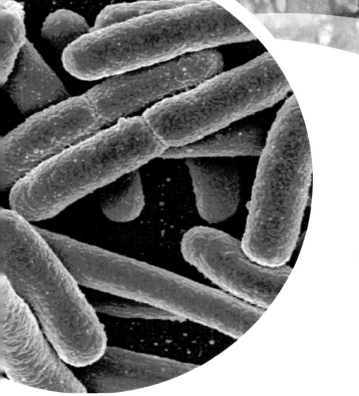

CONTENTS

chapter one

It's Alive!

WHAT'S THE difference between the flowers growing in your yard and the rocks surrounding them? What makes the bear at the zoo different from the stuffed one that sits on your bed? Hopefully, you recognize that the difference is that the flowers and the bear are living and the rocks and the teddy bear aren't. But what actually makes something "alive"?

IS IT ALIVE?

Look around you. What's alive and what's not? How can you tell? Is it because something moves? But lots of nonliving things move—water, machines, and air, for instance. On the other hand, a cactus doesn't seem to move, but it is very much alive. Can you tell something is alive because it's growing? But crystals grow, and they aren't alive. If something dies, does that mean it was living? But cars "die," and they weren't living.

So what *does* make something "living"? Biologists have developed a list of characteristics that **organisms** share. *All* living things do *all* of these things. Nonliving things might do one or more of them, but not all of them.

All living things have the ability to acquire and use energy, **reproduce**, respond to the outside world, and get rid of waste. Some biologists also add movement, growth, and a **life span** to the list. So while the clock on the wall might have hands that move and will eventually die (stop working), it doesn't change size, eat, have baby clocks, shiver from the cold, or dash to the restroom every few hours.

A Bit About Biologists

Biologists are scientists who study living things. Some biologists specialize in certain areas. A microbiologist studies microbes, which are tiny organisms (like cells!) that can only be seen with a microscope.

The last characteristic that all living things share is **cells**. All living things are made of cells. Some living things are made of only one cell. Other living things are made of millions of cells.

WHAT IS A CELL?

• • •

Cells are the building blocks of life. They grow. They react to changes in the environment. They change **nutrients** into energy. They reproduce. Cells are the smallest living units that can carry out these basic functions of life.

There are many different kinds of cells. The cells of a cucumber are not the same as the cells of a baboon. **Bacteria** cells do not have a **nucleus** like plant and animal cells do. But despite the differences, all cells are living and all living things have cells.

A Roomful of Cells

The word *cell* means "little room." Cells are like little rooms that hold important things an organism needs.

A microscopic picture of the bacteria that cause the disease anthrax (Bacillus anthracis)

Bone cell

Liver cells

Fat cells

The Scoop on Cells

THE DISCOVERY OF CELLS

● ● ●

MICROSCOPES WERE invented around 1660 by a Dutchman named Anton van Leeuwenhoek. By looking under his microscope, van Leeuwenhoek discovered an invisible world that no one had seen before. One of his discoveries was tiny "beasties" in pond water, blood, and even the scum he scraped off his teeth. (Later, these "beasties" were identified as cells.)

Anton van Leeuwenhoek

When the word got out about all the amazing things van Leeuwenhoek could see, other people began making and using their own microscopes. An English scientist named Robert Hooke looked at a slice of cork under his homemade microscope. The outline of "beasties" that he saw reminded him of the tiny rooms in a monastery. So Hooke called the shapes *cells*, which is what the rooms in the monastery were called. The name stuck, and from then on, the tiny building blocks of life became known as cells.

THE SIZE OF CELLS

• • •

Most cells are much smaller than the period at the end of this sentence. In fact, most plant and animal cells are between 10 and 50 micrometers wide. What's a micrometer? It's 1/1000 of a millimeter. If you look at a metric ruler, you'll see that a millimeter is about the width of a pencil line. Try to imagine one thousandth of that!

Try This!

Check out your cheek cells. Rub a toothpick gently against the inside of your cheek. Smear the stuff you scraped off your cheek onto a microscope slide. Put a drop of iodine on the smear to stain it so you can see the cells better. Add a cover slip. Look at the slide under the microscope. You should see some oval-shaped cells. Those are your cheek cells.

The world is full of tiny, single-celled organisms. You can't see them when you look around, but they're there. Right now, your skin is swarming with microscopic living things. These tiny organisms are swirling around in the air you breathe. If you put pond water under a microscope, you'll see all kinds of tiny creatures swimming around.

Calling All Single Cells!

Single-celled organisms are also called *unicellular organisms. Uni* means "one," so a unicellular organism is made of one cell.

The larger living things that you *can* see are made of huge numbers of cells. These bigger organisms, such as whales, trees, and elephants, don't have larger cells. Their cells are still microscopic. These bigger plants and animals just have a lot more of these tiny units.

Living things with many cells usually have different kinds of cells with special jobs. Your body has about 200 different kinds of cells. Each of these cells has its own job, but together they perform tasks in your body. Plant cells work this way too. Cells in the roots, stem, and leaves team up to help the plant breathe, reproduce, and make food.

Connective tissue cells

Smooth muscle cells

Skeletal muscle cells

THE SHAPE OF CELLS

Cells come in many shapes. They may look like cubes, boxes, rods, discs, spirals, or blobs. Many single-celled organisms are shaped like spheres (balls). Most plants with more than one cell have cubed or rectangular cells. Animals (including humans) have the greatest variety in the shapes of their cells.

CELL DIVISION

No, this is not a type of math. It's the way that cells reproduce. Each living thing starts life as a single cell. This first cell grows to a certain size and then splits into two cells. Later those two cells split again. More and more cells continue to form this way. As the number of cells increases, the organism gets bigger. By the time it stops growing, it may have trillions of cells.

chapter three

What's Inside a Cell?

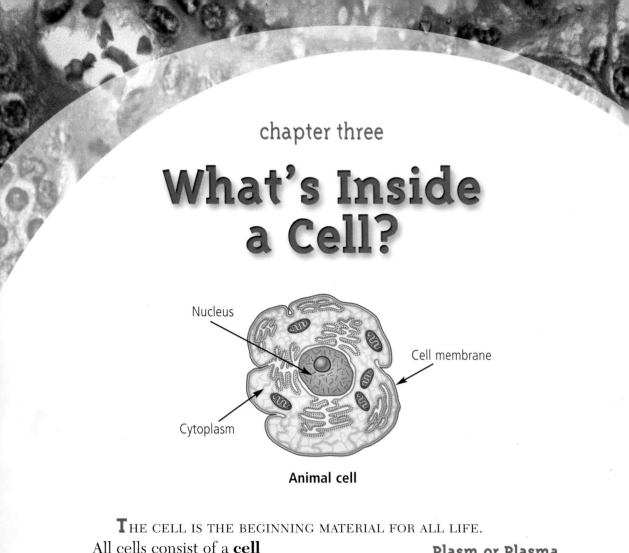

Nucleus

Cell membrane

Cytoplasm

Animal cell

THE CELL IS THE BEGINNING MATERIAL FOR ALL LIFE. All cells consist of a **cell membrane** that is filled with **cytoplasm**. Inside the cells of most organisms is a major structure called the *nucleus*. In addition, plant cells have a **cell wall** and **chloroplasts**.

Plasm or Plasma

Plasm and *plasma* are scientific words for the liquid part of substances that consist of a liquid and solid parts floating in it. For example, blood is plasma with red blood cells, white blood cells, and platelets floating in it. Cytoplasm is a liquid in cells that holds many tiny parts of the cell.

THE CELL MEMBRANE

● ● ●

Each cell has a "wrapper" around it called a *cell membrane*. The cell membrane is a thin, flexible, outer "skin" that controls the movement of materials in and out of the cell. The membrane has tiny openings that allow certain materials to enter and leave the cell. In human cells, the cell membranes allow useful substances such as oxygen and nutrients to enter and harmful wastes such as carbon dioxide to leave.

In some cells, the membrane has many wrinkles and folds. The additional surface area allows more material to pass in and out.

The cell membrane also holds the cell together and gives the cell its shape. Membranes are stretchy and flexible, so the cell can change shape when under pressure. Squeeze or press a part of your arm. Once you let go, the skin and muscles bounce back to the way they were. Thank your flexible cell membranes for that. If a membrane is stretched too far and tears, it can repair itself. What a handy thing to have around (your cells)!

Also Known as the Plasma Membrane

Cell membranes have characteristics that are similar to liquids. For example, a cell membrane changes shape when it's squeezed just like a liquid does when it's poured into a container. Because of this, scientists often refer to the cell membrane as the plasma membrane.

THE NUCLEUS

• • •

The nucleus is the command center of the cell. A nucleus is round or oval-shaped and is wrapped in its own membrane. It is usually found near the center of the cell. The nucleus of a cell is the "boss." It sends messages to the other parts of the cell telling them what to do. The nucleus controls all of a cell's activities, including how new cells are made. When scientists removed nuclei from living cells, these cells could not divide into new cells.

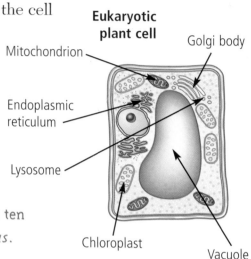

Eukaryotic plant cell

Mitochondrion

Golgi body

Endoplasmic reticulum

Lysosome

Chloroplast

Vacuole

Us or I?

One cell has a nucle**us**. Ten cells have ten nucl**ei**. *Nuclei* is the plural form of *nucleus*.

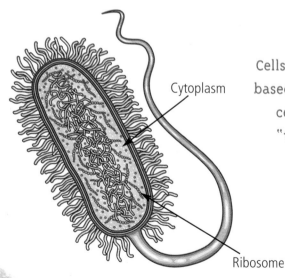

Cytoplasm

Ribosome

Prokaryotic bacteria cell

With or Without a Nucleus

Cells are divided into two main categories based on their inside structure. Eukaryotic cells have a nucleus. *Eukaryotic* means "true nucleus." Plants, animals, **fungi**, and **protists** have eukaryotic cells.

A prokaryotic cell does not have a nucleus. *Prokaryotic* means "before the nucleus." Bacteria cells are prokaryotic.

THE CYTOPLASM

• • •

Inside the cell membrane is the cytoplasm. Cytoplasm is the name for all of a cell's living material except the nucleus. *Cyto* means "cell," so *cytoplasm* means "cell stuff."

Cytoplasm is made mostly of water. In fact, cells are 70 to 80 percent water. Cytoplasm is a **translucent**, grayish, jellylike substance that looks somewhat like Jell-O. But unlike Jell-O, the cytoplasm flows within the cell membrane.

The cytoplasm holds dissolved nutrients such as proteins, which are the cell's building blocks. It also contains sugars, fats, and starches that provide energy. Salts, minerals, and vitamins also move around in the cytoplasm.

The watery gel of the cytoplasm is crowded with **organelles**. An organelle is a small part of a cell with a special job. **Mitochondria**, **vacuoles**, **endoplasmic reticulum**, **Golgi bodies**, and **lysosomes** are organelles.

Tiny Parts

The word *organelle* means "little organ." *Organ* is another word for "body part." The heart, lungs, liver, and kidneys are examples of organs. Organelles are little parts of cells.

Mitochondria

The mighty mitochondria are shaped like tiny rods or beans. Some cells have more than a thousand mitochondria. Mitochondria are like little power plants. They release energy by using oxygen to burn the fuel that comes from food.

Many Mitochondria

One little power plant in a cell is a *mitochondrion*. More than one of these organelles are *mitochondria*.

While they create energy, mitochondria also give off wastes, such as water and carbon dioxide. If a cell needs more energy, its mitochondria can grow and divide to make new mitochondria.

Mitochondria

Vacuoles

Vacuoles are a cell's storage organelles. They are like pockets or holes inside the cell that store water and nutrients until the cell is ready to use them. Vacuoles also store waste until the cell is ready to get rid of it. These organelles vary in size and can be almost as large as the cells themselves.

Endoplasmic Reticulum

The endoplasmic reticulum is a network of tubelike passageways that extend from the nucleus out into the cytoplasm. One of the main jobs of the endoplasmic reticulum is to help make and transport proteins. Proteins are structures needed for communication, growth, repair, and replacement of cells.

Dotted along the inner branches of the endoplasmic reticulum are tiny organelles known as **ribosomes**. Ribosomes make proteins and drop them into the reticular passageways. The tubes help move the proteins to other parts of the cell.

Not all ribosomes are found on the endoplasmic reticulum though. Some float around free in the cytoplasm.

Golgi Bodies

A Golgi body is a stack of flat, pancake-shaped discs. These organelles modify, sort, package, and store carbohydrates and proteins until they are needed. Then they ship out the materials to other locations, both in and out of the cell. Golgi bodies are also responsible for making lysosomes.

An Organelle with Other Names

A Golgi body is also called a *Golgi complex* or *Golgi apparatus*.

Lysosomes

Lysosomes are round organelles that digest cell nutrients. They are also the garbage collectors of the cell. They break down products that might harm the cell. They contain chemicals that help white blood cells break down harmful germs that have entered the body.

Try This!

Make a model of a cell. Gather two baggies with a seal, a cup of light corn syrup, and a variety of small foods (cereals, gummi worms, jelly beans, gum balls, etc.) to represent the organelles. Pour the corn syrup into one of the baggies. Place the filled baggie inside the other baggie. (This is just for added strength.) The baggies are the cell membrane, and the syrup is the cytoplasm. Place a round item in the middle of the syrup for the nucleus. Then put all of your other organelles in the baggies and seal. On a sheet of paper, list each food item and which organelle it represents. Display your cell and labels.

PLANT CELLS
● ● ●

Both plant and animal cells have a cell membrane, a nucleus, and cytoplasm with organelles. Plant cells, however, also have chloroplasts and cell walls.

A human's cells can't make their own food, but a daisy's cells can. Plant cells contain tiny green food-making organelles called *chloroplasts*. Chloroplasts have a green material in them called *chlorophyll*. Chlorophyll takes energy from the Sun and uses it to help turn carbon dioxide and water into sugar that the plant can use as fuel. This process is called *photosynthesis*, which means "putting together by light."

The other big difference between plant and animal cells is the cell wall. Plant cells have a cell membrane around the outside just like an animal cell has. But a plant cell also has a cell wall outside of the membrane. This wall is made of a material called *cellulose*.

Cellulose is made of long chains of sugar **molecules** that the cell makes.

The cell wall is thicker and tougher than the cell membrane. This wall protects and supports the plant. But while the wall is stiff, it does let some materials, such as water, carbon dioxide, and oxygen, pass through it.

Fiber Facts

It is the cell wall that gives fresh vegetables their crunch when you bite into them. This wall also provides you with fiber. Fiber is necessary for good digestion. Your body can't break it down into usable food, so it passes right through you. This keeps your digestive tract clean and healthy.

chapter four

The Secret Life of Cells

SO WHAT DO CELLS DO ALL DAY long? Sit around and talk on cell phones? No, they're much too busy for that. Doing what? Lots of things!

Cells take in nutrients, which they use to provide energy. They make proteins and get rid of wastes. They grow and divide, producing more cells. In multicellular organisms, each cell is not only responsible for itself but for contributing to the life of the whole organism as well.

Many Meanings

If you are wearing a multicolored shirt, what color is your shirt? *Multi* means "many." A multicolored shirt is many colors. Therefore, *multicellular* means "having many cells."

A CELL'S DIET
• • •

When a cell needs food, does it just gobble up a slice of pizza? No, you do that for your cells. Cells get the nutrients they need from the food you eat. The tiny but mighty mitochondria combine the food you eat with oxygen and turn it into the materials your cells need to do their jobs.

DIFFUSION
• • •

Mitochondria need food (fuel), oxygen, and water to make energy. They put out carbon dioxide as a waste product. But how do these materials move in and out of the cells? **Diffusion** is one process that makes this possible.

The molecules, or tiny particles, of a substance are constantly moving. This motion causes them to spread out. In diffusion, tiny molecules of a substance move from a place where they are crowded together to a place where they are less crowded.

The movement of oxygen from red blood cells to other body cells is an example of diffusion. Red blood cells carrying a large amount of oxygen pass nearby cells with little or no oxygen. The oxygen then diffuses out of the red blood cells into the other cells. Meanwhile, carbon dioxide particles are crowded in the body cells, so they diffuse to the blood.

Try This!

See diffusion in action with some food coloring, watercolor paint, or ink. Fill a clear glass or jar with water. Add a drop of food coloring, paint, or ink. What happens to the color? The crowded color molecules spread out in the water.

You can also smell diffusion in action. Spray some perfume or air deodorizer in a room with several people in it. Ask each person to raise his or her hand when first smelling the scent. Who can smell the diffused scent first?

In order for a person to smell something, molecules of that substance must diffuse through the air. Some of them travel into the person's nose, where the scent is recognized by the brain. This happens even when the air is still because the molecules move from the crowded sprayed area to the less crowded air.

OSMOSIS

• • •

Water is the most important substance that enters a cell. Most of the water that cells get comes in through **osmosis**. Osmosis is the diffusion of water through cell membranes.

It's easy to observe the results of osmosis in a plant. When there is more water in the soil than in a plant's roots, the water will diffuse into the root's cells. Water then enters the other plant cells and gets stored in their vacuoles. This plumps the cells up so that their cell membranes push against the cell walls. The plant stands up and looks perky. If the soil becomes drier than the roots, then water starts leaving the plant by osmosis. When cells lose water, they shrink. This makes the cell membrane shrink away from the cell walls. This causes the plant to wilt.

Try This!

Check out osmosis in french fries. Gather two dishes of tap water, some salt, and a sliced potato.

Make salt water in one dish by stirring in salt until it won't dissolve anymore. Don't add salt to the other dish of water. Place half of the potato slices in each dish. Let them sit for at least an hour. What happens to the potatoes?

The potato pieces in the salt water will be limp. The water in the potato moved into the salt water. The potato pieces in plain water should look the same as they did when you put them in.

Why are the two different? Potato cells contain a lot of water. The salt in the dish spread the water molecules apart. Since the water molecules were packed tighter in the potato than in the salt water, the water molecules moved to where they were less crowded.

CELL DIVISION
• • •

Cells reproduce themselves. That means they make new cells. These new cells are made when an existing cell divides into two new cells. This is called *cell division*.

Why do cells divide? Cells divide for different reasons. The cells of unicellular organisms divide for only one reason—to maintain their population. If they didn't make new cells to replace dying ones, eventually they would all disappear. The reproduction process of these organisms involves a single cell splitting into two new identical cells. This process is called *fission*.

A Family of Cells

When cells reproduce, the original cell is called the *parent cell*. The two new cells are called *daughter cells*.

The cells of multicellular eukaryotic organisms, such as plants and animals, divide for several reasons. These types of organisms use cell division to grow, to replace old and worn-out cells, to repair areas of the body that have been damaged, and to produce **offspring**.

To grow, replace, and repair, the cells produce daughter cells that are exact copies of the parent cell. This process is known as **mitosis**. It is similar to the process of fission. In mitosis, the nucleus of the parent cell divides into two identical nuclei in the daughter cells.

Inside a cell's nucleus are thread-shaped chromosomes. Chromosomes direct all of the activities in a cell. Each chromosome may have thousands of genes. A gene is an instruction for a characteristic of a cell. Genes are made of deoxyribonucleic acid (DNA). This DNA passes on characteristics, such as height or eye color, from an organism to its offspring. Luckily, before a cell divides, the nucleus makes a complete copy of all the chromosomes—one for each daughter cell.

Human embryo

So why aren't you an exact copy of your parents? This happens (or doesn't happen) because multicellular organisms use a different type of reproduction called *sexual reproduction*. This process requires a different type of cell division. Sexual reproduction produces offspring that are similar but not identical to the parents.

The cells used for sexual reproduction are called *gametes*. They are formed by a type of cell division that creates cells that contain half of the chromosomes an offspring will have. This process is known as **meiosis**. When the mother and father gametes join together, the offspring now has a complete set of chromosomes. Because the offspring inherits traits from both parents, it is different from both of them. It is a unique, one-of-a-kind organism.

How Many Chromosomes?

Each type of organism has a different number of chromosomes. Humans have 23 pairs of chromosomes. Frogs have 13 pairs. Pea plants have only 7 pairs of chromosomes.

A TIME TO DIVIDE
AND A TIME TO DIE
● ● ●

Cells don't divide until they have grown to their full size. The new daughter cells won't divide again until they are "grown up."

How long does it take a cell to divide? Some cells take less than an hour. Many take 12 or more hours. It depends on the kind of cell. Animal cells finish their division process by pinching in two. Plant cells finish by building a new cell wall. Whether it takes a few minutes or a few hours, each type of cell has its own remarkable process of reproducing itself.

Like other living things, cells grow old and die. Different types of cells have different life spans. Liver cells live about 18 months. Red blood cells live about 2–4 months. Skin cells have a life span of about 20 days. Cells in the lining of the stomach and intestines have short life spans of only 2–5 days. Since all of these cells can reproduce themselves, the length of their life isn't really important.

The nerve cells in a human are an exception to this. Nerve cells don't reproduce. When these cells in the brain, spinal cord, and nerves die, they can't be replaced. Luckily, however, nerve cells can live more than a hundred years.

Life and Death Statistics

● Every minute, approximately 3 billion cells in your body die. Of course, in that same minute, 3 billion cells are created through mitosis.

● Scientists estimate that as much as 2 percent, or 1 out of 50, of the cells in your body die every day.

● If you are still growing, more cells are made than lost in your body each day.

chapter five

A Team of Cells

Protozoa

MOST LIFE-FORMS ARE MADE OF a single cell that does all the things needed to stay alive. Bacteria and **protozoa** are two common single-celled organisms. These tiny organisms are normally so small that they can't be seen with the human eye. So until the microscope was invented, nobody knew they existed. Now we know that these organisms live almost everywhere—even in and on our bodies.

In multicellular organisms, such as mushrooms, marigolds, mice, and men, cells don't live on their own. They work together with other cells to form a large organism. The survival of these organisms depends on all of the cells working together properly.

This group approach by cells has many advantages. It enables an organism to grow larger. It also allows the cells to divide up the workload. It takes a lot of energy to keep a large organism alive and healthy. No cell can do it on its own.

It's kind of like sharing the household chores at your home. Imagine if you had to do all the cooking, cleaning, laundry, etc. Instead, your family divides up the jobs. Perhaps it's your job to take out the garbage and vacuum the carpet, while your sister dusts and sets the table. Just like in a household, cells share the work in a multicellular organism.

SPECIALIZED CELLS

Different types of cells specialize in specific jobs. In plants, for instance, some cells carry nutrients, while others use sunlight to make food. Some animal cells control movement, some carry nerve messages, and others store energy. Some of these specialized cells in the human body are blood cells, skin cells, nerve cells, muscle cells, and fat cells.

Blood Cells

There are three different kinds of blood cells—red blood cells, white blood cells, and platelets. Humans have billions of red blood cells. These cells carry oxygen to cells and carbon dioxide away from cells. White blood cells battle harmful germ invaders. Platelets are actually pieces of cells. They help blood to clot, or stop flowing, when necessary.

Red blood cells on an artery wall

Skin Cells

Skin is the largest and fastest-growing organ. The billions of skin cells that form skin weigh about seven pounds. Skin cells stack together like millions of tiny bricks to make a protective "wall" around your body.

The skin's outer layer is called the *epidermis*. This layer is mostly the remains of dead skin cells. Millions of these dead cells flake off every day.

Underneath the epidermis is the dermis layer. This layer of the skin is strong and thick and performs many duties.

The dermis allows you to sweat. The nerves in this layer help you feel objects. Cells in the dermis also secrete oil that keeps skin soft and waterproof.

Human skin cells

Epidermis

Dermis

Try This!

Scratch your scalp, and then check under your fingernails. Did you collect any dead skin cells? Or use a pumice stone to rub the heel of your foot. Did you rub off any dead skin cells? Gather a few of these cells, and check them out under a microscope. Even dead, these cells are interesting to observe.

Nerve Cells

Nerve cells are found in the brain, spinal cord, and nerves. Half of the ten billion or more nerve cells in a human body are in the cerebrum. That is the part of the brain that thinks. You are using it right now to read this book.

A nerve cell is shaped a bit like a spider with a long strand of spider silk attached to it. This long strand is called an *axon*. Axons carry nerve impulses, or messages, from one nerve cell to the next. The body of a nerve cell has fibers sticking out of it called *dendrites*. Incoming messages reach the cell through these tiny branches. Outgoing messages travel from the cell body through the long, thin axon. Some axons run from the spinal cord in the back all the way down to the foot.

Nerve cells are made to last a lifetime. They do not reproduce. This is why it's important not to drink alcohol, take drugs, or ride a bike or skateboard without a helmet. Once a brain cell is damaged or destroyed, it's gone for good.

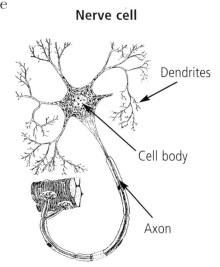

Nerve cell

Dendrites

Cell body

Axon

Touchdown!

Messages travel through nerve cells at about 70 miles per hour. That means they move about the length of a football field each second.

Cardiac
muscle cells

Muscle Cells

Muscle **tissue** is made of muscle cells. Muscle cells are the only kind of cell in the body that works by contracting (shortening) and relaxing (lengthening). There are three kinds of muscle cells—cardiac, smooth, and skeletal.

Cardiac muscle is found only in the heart. Muscle cells in the heart contract, causing the heart to pump blood to the body. These cells relax between heartbeats, allowing blood to enter the heart. Cardiac muscle contracts and relaxes involuntarily. This means the cells do it on their own without the person thinking about it. Imagine how hard it would be if you had to think about every beat of your heart!

Smooth muscles are found in body organs, such as the stomach and intestines. They are called *smooth* because they lie in flat sheets against the walls of the organs. These cells also contract and relax involuntarily.

Most muscle cells are skeletal. Skeletal muscles are also called *striated muscles* because the insides of the cells have bands of light and dark striping. The main job of skeletal muscles is to pull on bones to make them move. Skeletal muscle cells usually move voluntarily. When you want to kick a ball, you command the muscles in your leg to move. Skeletal muscles can move involuntarily too. When your hand jerks away from a hot stove, this is an automatic response that you don't control.

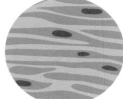

Smooth
muscle cells

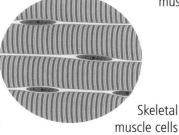

Skeletal
muscle cells

Getting in Shape

The job that a cell does often determines its shape. Red blood cells are shaped like jelly donuts with a dented center. The thin center allows them to bend when squeezing through tiny blood vessels. Nerve cells are long and threadlike. These thin "branches" stretch out through the body to deliver and receive messages from the brain and body. Muscle cells are long and thin so they can stretch and shorten to cause movement.

Fat Cells

Groups of fat cells look a bit like bubble wrap or honeycombs under a microscope. Individual cells look like slices of banana. The peel is the small cytoplasm and nucleus. The fruit is the fat droplet.

Fat cells provide a layer of insulation to help keep a body warm. These cells also store energy for later use. Because most fat lies just below the skin, it serves as a padding to protect deeper tissues and organs from injury.

If you store more fat than your body needs, you gain weight. However, when you gain weight, you don't produce more fat cells. The cells you have just get bigger.

Brown Baby Fat

Most of the fat in an adult's body is white fat. Babies, however, are born with brown fat. This brown fat makes heat for the baby since he or she hasn't had time to form much white fat for insulation. As the baby develops more white fat cells, the brown cells disappear.

THE HUMAN TEAM
● ● ●

The cells in the human body are arranged in a highly organized way. Similar cells working together to do a special job are called *tissue*. Humans and animals have four types of tissue—muscle tissue, nerve tissue, connective tissue, and epithelial tissue. Most body tissue is muscle tissue. Nerve tissue is found in the sense organs, brain, spinal cord, and nerves. The skeleton uses connective tissue. Blood is also considered connective tissue. Epithelial tissue includes the skin and the lining of most organs.

Let's Call It Tissue!

An English botanist, N. Grew, was the first to label groups of cells working together. He called them *tissues* because he noticed that they were often in the form of thin sheets.

Tissues working together form organs. Organs are groups of different tissues working together to perform a specific function in the body. The brain, stomach, heart, and lungs are organs. Each organ is made of more than one kind of tissue. A heart, for instance, is made of muscle tissue, blood tissue, and nerve tissue. Each of these tissues plays its own role in the heart so it can pump blood to the entire body.

Organs that work together are called a *system*. The human body has several systems. The skeletal system (bones) protects and supports the body. The muscular system (muscles) supports and moves the body. The digestive system (esophagus, stomach, and intestines) breaks down food and changes it into nutrients that the body needs. The circulatory system (heart, blood, and blood vessels) moves oxygen, wastes, and digested food throughout the body. The nervous system (brain, spinal cord, nerves, and sense organs) acts as a command center that conducts messages throughout the body. The respiratory system (lungs, windpipe, nose, and mouth) takes in oxygen and gets rid of carbon dioxide. Several other systems are also hard at work in the human body.

Plants Have Organs Too!

Like animals, plants have organs too. Roots, stems, leaves, and flowers are all organs in a plant. Each organ has a specific job, such as collecting water, making food, and reproducing.

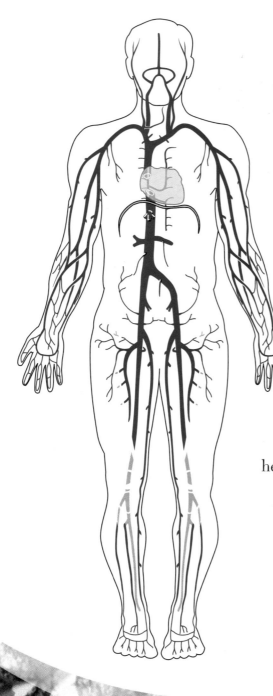

TEAMWORK

• • •

From individual cells to tissues to organs to systems, an organism is a complicated arrangement of cells. Each cell has its own position to play on the body's team. Working together, the trillions of cells in the human body are tiny players that strive to keep the person alive and healthy. Go team!

chapter six

Good Guys and Bad Guys

Most cells in a multicellular organism have a positive purpose. They produce energy, distribute nutrients, get rid of wastes, or perform other important jobs. These cells are the "good guys" that every organism depends on.

However, there are a few "bad guys" that can damage an organism's system, causing illness and even death. Understanding these invaders can help protect you against their dangers.

WHITE BLOOD CELLS

● ● ●

White blood cells are the superheroes of the cell world. These tiny guys are fighting for you and dying for you right now. White blood cells patrol your bloodstream looking for germs. When "bad" bacteria invade your body and cause an infection, white blood cells rush in to gobble them up. These cells can squeeze through the walls of tiny blood vessels to get to an infection. Once they arrive, they use one of many ways to immobilize the invader. One of these ways is to wrap themselves around the bacteria and "eat" them.

Eventually, all white blood cells die. Some live for only a few hours. Others last for months or even years. Many die battling the "bad guys." The thick yellowish pus found in infected cuts or in your nose when you have a cold is a sign of their death. Pus is a mixture of dead bacteria, white blood cells, and other body cells. The invisible army of white blood cells loses defenders every day protecting the rest of your cells.

Color Confusion

White blood cells aren't really white. They are colorless.

BACTERIA

• • •

Bacteria are everywhere. They live in dirt, air, and water. They live in and on your body. In fact, billions and billions of bacteria are living happily in your mouth right now. Your bad breath and underarm odor can be blamed on bacteria too. So what are all these bacteria doing?

Bacteria love to multiply when living conditions are good. They thrive in environments that are warm, dark, and moist. If things are just right, each bacterium can split in two every twenty minutes. If those two split again, you have four, then eight, then sixteen, then thirty-two, and so on. Within a few hours, one bacterium can turn into millions. That's a lot of bacteria!

Staphylococcus aureaus bacteria can cause skin and wound infections, pneumonia, and food poisoning.

Um . . . Ah . . . How Many?

A single bacterium multiplies into many bacteria. *Bacteria* is the plural form of *bacterium*.

Ancient Cells

Bacteria are the oldest known living things. There is fossil evidence that these cells have been around three and a half billion years.

Bacteria cells are both good guys and bad guys. Helpful bacteria far outnumber the harmful ones. Only about 1 out of 30,000 types of bacteria are actually harmful to other organisms.

Many bacteria are useful. Bacteria are used to make dairy products, such as sour cream, yogurt, and cheese. Bacteria put the "sour" in sourdough bread. The bacteria in your stomach help digest food. Some bacteria are used to make antibiotics, which are medicines that kill harmful bacteria. Bacteria in soil take nitrogen gas from the air and turn it into materials that help plants grow.

Bacteria also play a big role in helping decompose, or break down, dead things. When a dead plant or animal decomposes, it returns to the soil where plants can use it to help them grow. If not for bacteria and other decomposers, we would have dead things piled up to the sky.

A few types of bacteria can cause trouble. If "bad" bacteria get on food and the food is left in conditions that encourage bacteria to multiply, the bacteria can cause food poisoning. That's why it's important to keep cold foods in the refrigerator.

Some bacteria make plants and animals sick. Gardens wither under attack from bacteria. Pets need a trip to the vet when bacteria visits. Farmers lose crops and cattle when bacteria strikes.

Bacteria in the human body are responsible for many "yucky" things. If you don't keep your teeth clean, bacteria will sit on them and cause cavities. Your skin is home to swarms of invisible bacteria. You can have up to about two and a half million bacteria on a square centimeter of skin. If those bacteria enter a cut or scrape, they can cause an infection. Some illnesses, such as strep throat, ear infections, and pneumonia, are caused by bacteria.

An Excuse for Messy Rooms

Scientists have learned that kids raised in superclean houses are more likely to get sick when exposed to bacteria. Exposure to bacteria when growing up allows kids to develop healthy immune systems that help fight off harmful invaders. It also reduces the occurrence of allergies. So tell that to your parents the next time they tell you to clean your room!

Watch Out for Germs!

We talk about germs all the time. *Cover your mouth when you sneeze so you don't spread your germs. Wash your hands to get rid of the germs.* How many times have you heard these statements? But what are these germs?

Germs are tiny invaders that can cause disease. The four major types of germs are bacteria, viruses, fungi, and protozoa. Each one can invade other organisms and make them sick. That's why we're always trying to get rid of them!

PROTISTS

Protists are one of those microscopic organisms running around everywhere. Protists are divided into three groups based on their characteristics. They either behave like plants, animals, or fungi. Many protists live in water. The algae (seaweed) you see when swimming in a lake or on rocks along the shore are protists.

Most protists are harmless, but a few of them can be destructive. Some protists harm fish and feed on their body fluids. Others use up large amounts of oxygen to decompose, which can be harmful to animals and plants in the area. A few protists are responsible for causing diseases, such as dysentery and malaria, in humans.

Take a water sample from a nearby pond, lake, or river. Place a few drops of the water on a microscope slide and observe. Do you see any protists (or other microscopic organisms) moving around?

VIRUSES

• • •

Viruses are strange things caught between the land of the living and nonliving. They are not considered living things because they cannot carry out any life processes on their own. They exist inside the cells of other organisms. There, the virus is able to reproduce. Viruses can invade a cell, take control of it, and make the cell produce new viruses. Once inside a body, viruses spread and can make a person sick. Colds, the flu, and chicken pox are all viruses.

Poisonous Invader

The word *virus* comes from a Latin word meaning "poison."

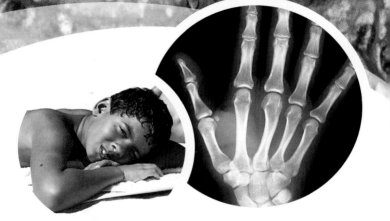

CANCER CELLS

• • •

Unlike healthy cells, cancer cells divide too often. Normal cells reproduce only enough cells to replace those that have died. Cancer cells multiply too quickly, crowding out normal cells and destroying healthy tissue. A buildup of cancer cells forms a tumor, or lump. The tumor eventually damages or destroys all the healthy tissue surrounding it. Cells can even escape from the tumor and spread throughout the body, causing more damage.

All of the causes of cancer are not clear, but it *is* known that some substances may contribute to the development of cancerous cells. Tobacco, some chemicals, and overexposure to sunlight and X rays may increase the risk of getting cancer. Some viruses may even trigger it.

Doctors treat cancer in three main ways. Surgery is done to remove cancer cells. Chemotherapy uses drugs to kill cancer cells. Radiation treatment uses X rays to destroy cancer cells.

• • • • • •

Whether they are good guys or bad guys, cells are the basic unit of life. Every living thing on Earth is made of cells. Without them, there would be no life.

It's hard to imagine that before the invention of the microscope, no one knew that cells existed. Today, powerful microscopes continue to reveal more about the wonderful workings of cells. Scientists continue to try to learn their secrets. As they learn more about cells, they learn more about life itself. Isn't it amazing how important something so small can be!

INTERNET CONNECTIONS FOR CELLS

• • •

http://www.eurekascience.com/ICanDoThat/
Chloe the Chloroplast and her friends will tell you about bacteria and plant and animal cells.

http://www.cellsalive.com
This interactive site will bring cells alive for you. View realistic diagrams of the parts of a cell, and read about their jobs. Visit the gallery of cells to see different types of cells. Watch the mitosis of a cell and the cell cycle as it changes. Keep an eye on the cell cams that show how bacteria and cancer cells grow.

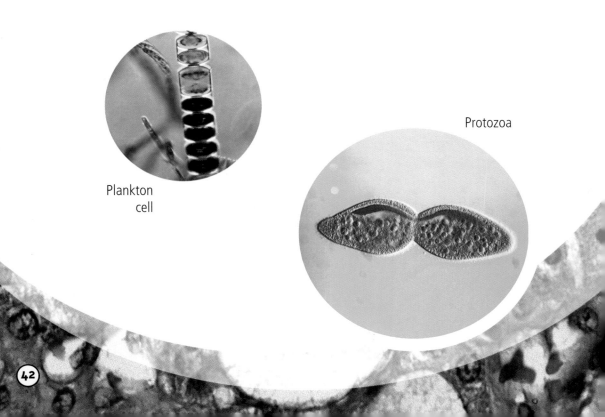

Plankton
cell

Protozoa

Algae cells

http://www.tvdsb.on.ca/westmin/science/sbi3a1/Cells/cells.htm
Check out colorful diagrams with information about the parts of a
cell and how they work.

http://waynesword.palomar.edu/lmexer1a.htm#n
Click on any part of the large plant and animals cells found here for
a description of its role in the cell's life.

http://www.microbe.org/microbes/what_is.asp
Solve microbe mysteries such as "Friend or Foe?" or "Virus or
Bacteria?" Bacteria, protists, fungi, and viruses won't be such a
mystery after you explore the information on this site.

http://www.gp4k.com/homeschool/Article.asp?ID=582
Read the story of Anton van Leeuwenhoek and his discovery of the
microscope and "beasties" (cells).

Glossary

bacteria   (bak TEAR ee uh) single-celled organism without a nucleus

cell   (sel) smallest unit of life

cell membrane   (sel MEM brayn) thin, flexible band around a cell that controls the movement of substances in and out of the cell

cell wall   (sel wawl) wall around a plant cell that protects and supports

chloroplast   (KLOHR uh plast) organelle that makes food for plants

cytoplasm   (SEYET uh plaz uhm) all of a cell's living material inside the cell membrane except the nucleus (see separate entry for *cell membrane*)

diffusion   (dif YOU zhun) process of molecules moving to a less crowded area (see separate entry for *molecule*)

endoplasmic reticulum   (EN doh plaz mik ruh TIK you luhm) tubelike passageways that extend from the nucleus into the cytoplasm (see separate entry for *cytoplasm*)

fungi   (FUHNG eye) organisms, such as mushrooms, that can't produce their own food

Golgi body   (GAWL jee BAH dee) organelle that packages and releases materials in a cell

life span   (leyef span) length of time an organism normally lives

lysosome (LEYE suh sohm) organelle that breaks down materials in a cell

meiosis (MEYE oh sis) process of dividing a nucleus that allows cells to produce gametes, or cells used for sexual reproduction

mitochondria (meye tuh KAHN dree uh) organelles that use oxygen to provide a cell with energy

mitosis (MEYE toh sis) process of dividing a nucleus that allows cells to divide into two identical cells

molecule (MAHL uh kyoul) tiny particle of a substance made up of two or more units of matter

nucleus (NOO klee uhs) organelle that controls a cell's activities

nutrient (NOO tree uhnt) material needed by living things to live and grow

offspring (AWF spring) new organism; child

organelle (OR guh nel) part of a cell found in the cytoplasm (see separate entry for *cytoplasm*)

organism (OR guh niz uhm) living thing

osmosis (AHZ moh sis) movement of water across a membrane

protist (PROH tist) type of organism with a nucleus

protozoa (proh toh ZOH uh) single-celled protists with animal-like qualities (see separate entry for *protist*)

reproduce (REE proh doos) to make more of something

ribosome (REYE buh sohm) organelle that makes proteins

tissue (TISH you) group of similar cells working together

translucent (TRANS loo sent) able to let light pass through

vacuole (VAK yuh wohl) organelle that stores materials in a cell

Index

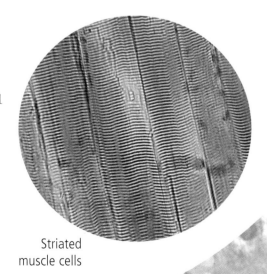

Striated
muscle cells

Chloroplasts inside
seaweed cells

Spirogyra
(freshwater algae)
cells